Theoretical

Free Electron Extraction Device (aka FEED)

Concept

by Dino Rondić

This book is dedicated to my loving parents.

Table of Contents

Chapter 1..7
Free Energy Concept..7
Chapter 2..10
The idea of extracting electrons from the electrically neutral
materials..10
Chapter 3..15
Basic Principles of Electron Extractor....................................15
      How this works?!...16
Chapter 3..22
Parts and 3D View...22
Conclusion...26

Chapter 1

Free Energy Concept

Throughout the history, there were many concepts of free energy, from the early concepts of overbalanced wheels which are made to be perpetual motion machines, to the modern solar and wind energy harvesting devices as are solar cells and wind turbines. Unfortunately, all of the devices proposed so far were unreliable and some completely nonfunctional.

There are three kinds of free energy devices. First kind is perpetual motion machines. Those machines were made to continuously move by them selves, without any external forced applied, and in the process they would give extra energy without interrupting their self-movement therefore creating eternal power source. Perpetual motion in an isolated system that violates the first law of thermodynamics and/or second law of thermodynamics. An Example of that kind of machine is the drawing of the overbalanced wheel of Leonardo da Vinci.

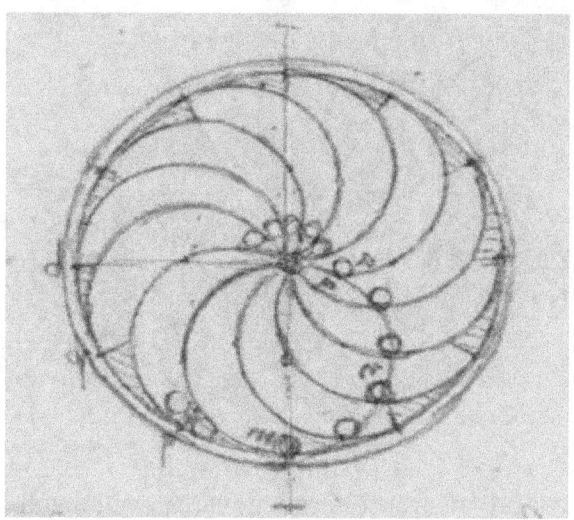

Illustration 1: 'Study in Perpetual Motion' (detail), Forster Codex, Volume II-2, 91v, 1495-97.

Although there are many more perpetual motion designs, we will not be addressing them in this book.

The second kind of free energy devices are the devices that extract energy from nature, more accurately, from Sun and atmospheric conditions that are indirectly related to Sun's and Moon activity. Those devices are solar cells, wind turbines, tidal waves turbines, etc. These devices and machines are very good in extracting energy from nature, although they are very much inconstant power sources due to their dependency on atmospheric conditions. Sun is not shining on the same region entire day, wind is not continuously blowing, tidal power is dependent on the tides, etc.

Illustration 2: Serpa Solar Park built in Portugal

This kind of free energy devices and machines is very much useful and necessary, but still, it is not meeting the criteria on perfect free energy devices.

Third kind of free energy devices, is the kind in which the subject of this book, the Electron Extractor, belongs. This kind of devices are dealing with the power on atomic level, electrons and protons.

The Electron Extractor is dealing with basic extraction of electrons from the materials that are in the electrically neutral state.

Chapter 2

The idea of extracting electrons from the electrically neutral materials

Every metal contains *free electrons.* Free electron is any electron that is not attached to an ion, atom, or molecule and is free to move under the influence of an applied electric or magnetic field.

Here is the list with concentration of free electrons in some metallic materials.

Metal	[1]Free Electrons per m^3
Copper	$8.47 * 10^{28}$
Lead	$13.2 * 10^{28}$
Zinc	$13.2 * 10^{28}$
Gold	$5.9 * 10^{28}$
Aluminum	$18.1 * 10^{28}$
Cadmium	$9.27 * 10^{28}$
Iron	$17 * 10^{28}$

When the force of the electric or magnetic field is applied to the conductor, these electrons flow through that conductor, which is called electrical current.

But to achieve the flow of free electrons, you need

---

[1] The charge of one electron is known to be about $1.602177*10^{-19}$ Coulombs, so 1 Coulomb is about $6.241509*10^{18}$ electrons = 1 Amp/sec. Since there is about $18.1*10^{28}$ electrons in a single cubic meter of aluminum, then there is about 8333333Ah of power in $1m^3$ of aluminum of free electrons stored. For the purpose of comparing, a common car battery, when fully charged, has a power of 45Ah stored in it.

electrical or magnetic field which is generated by using physical force (rotating the generator).

However, rotating the generator requires force which requires investment as fuel, or man power. We don't need that if we want to create free power device.

If we manage to create the device that is able to extract free electrons from the metal without investing any energy into the process, we then will be able to say, that that device if free energy device. But how we will manage to attract the free electrons to exit the material and go through the conductor so they can eventually end up "beating" the grid of the resistor and transferring the energy into the heat.

Now, lets address to the issue of the attracting the electrons so they can leave the host material and go through the conductor.

Basic idea behind this, is coming from the simple electrotechnical and physical laws of opposite charges attracting each other.

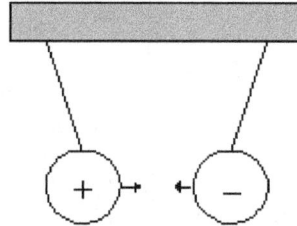
Illustration 3: Opposites Attract

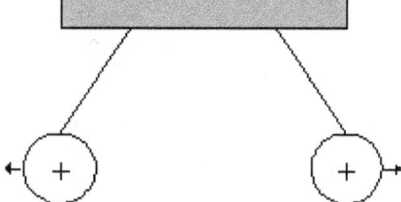
Illustration 4: Similar Repel

By knowing that opposites attract, we can now construct the basic principle on which this electron extractor works. We can take two material surfaces. One surface we will cal the Anode (positively charged electrode), and the other surface we will call the Inducted Charge Surface.

This works on the same principle as the above described "attraction" law. If we fill the Anode with positively charged particles (protons) and move the Anode close to the metallic surface which we already have named Inducted Charge Surface, as the natural response to the close proximity of those surfaces, the Inducted Charge Surface will fill up with the negatively charged particles (electrons). This happened because the positively charged surface, or the Anode, is acting as a magnet for the negatively charged particles in the other surface, so they emerge from the body of the metallic material to its surface.

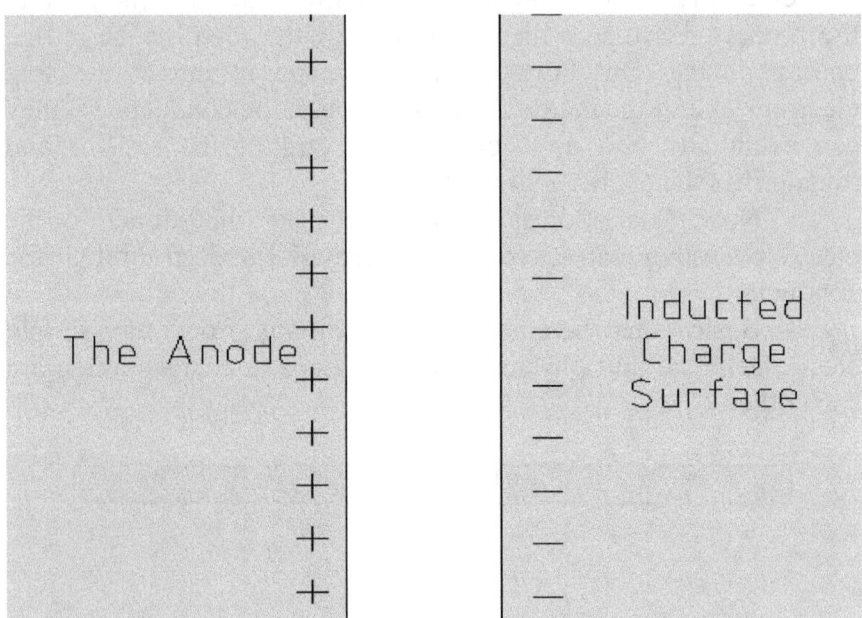

Illustration 5: Attracting the Electrons with placing Anode in close proximity to the neutral metallic material surface

To make all of this more practical, lets make this structure less theoretical. Lets attract the electrons from a metal ball to the surface. We will place the Anode to the one side of the ball, and the other side will be left without electrical force applied.

See the two-dimensional drawing bellow for the results of this experiment.

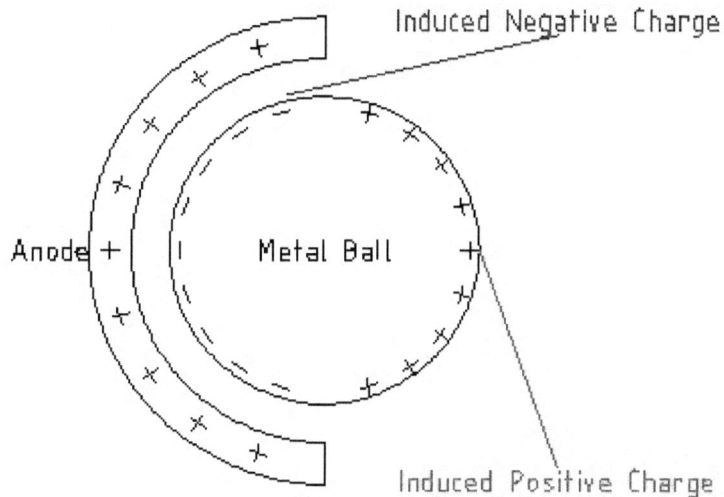

Illustration 6: Experiment with the metal ball

From the examples above, we are now getting the large picture. The result is now obvious, we need to attract the electrons from the metal to its surface so we can try to "gather" them.

Chapter 3

Basic Principles of Electron Extractor

We have learned from the previous chapter, how we can extract the free electrons from the metal. Now it is time for something more practical, we will learn how would the basic Electron Extractor machine work.

But before that, lets see some statistical data of all of this to drive away any doubts you might have on usability of this kind of device.

For example, in the 1 cubic meter of aluminum, or in the peace of aluminum the size of one meter by one meter by one meter, there is about $18.1*10^{28}$ of free electrons (see table on chapter 2).

Since 1C (Coulomb) is equal to the $6.241509*10^{18}$ of electrons.

$$1C = 6.241509*10^{18} \, e$$

And 1C is also equal to 1 ampere per second:

$$1C = 1 Ampere * 1 Second$$

Therefore:

$$6.241509*10^{18} \, e = 1 Ampere * 1 Second$$

Since there is $18.1*10^{28}$ e per m³ of aluminum, from that, we have:

$$18.1*10^{28} e / (6.241509*10^{18} \, e * 3600) = 8055388.1726 Ah$$

So, we have 8055388.1726Ah in one cubic meter of aluminum, that is equal to 179008.6 times more power than one common 45Ah car battery contains.

Now, lets back to the issue of explaining basic principles of free electron extractor.

Attraction of free electrons on the surface of metal ball is very interesting, however, it does not have any engineering value so far. If we could gather the electrons once they are on the surface of the metal ball, that would be of a value to us.

Maybe we can gather those electrons. I have developed a theoretical concept of using those attracted electrons. We can attract free electrons on the surface of the metal ball even if the Anode is not in the close proximity to the metal ball it self, but uses a mediator like an plate on the conductor.

This works on the simple process, expanding the Anode with an conductor as shown on the drawing below.

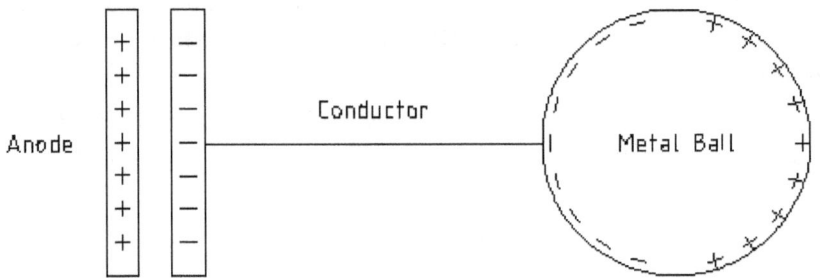

Illustration 7: Attraction of free electrons with an expanded Anode

Illustration 7 is an updated version of illustration 6, now the anode is not in the close proximity to the metal ball, yet it still does exactly the same thing as on the illustration 6.

The anode attracts the free electrons from the metal ball, then those electrons travel through the conductor to the another electrode.

How this works?!

Basic principle behind this free electron attraction is that, in this way, we create electric potential difference between the anode and the center of the ball. This is called Electric Voltage, and every regular electric generator works on the same principle.

Electric Voltage drives the electrons from the ball to the anode the same way as it drives the electrons basically in every generator known today.

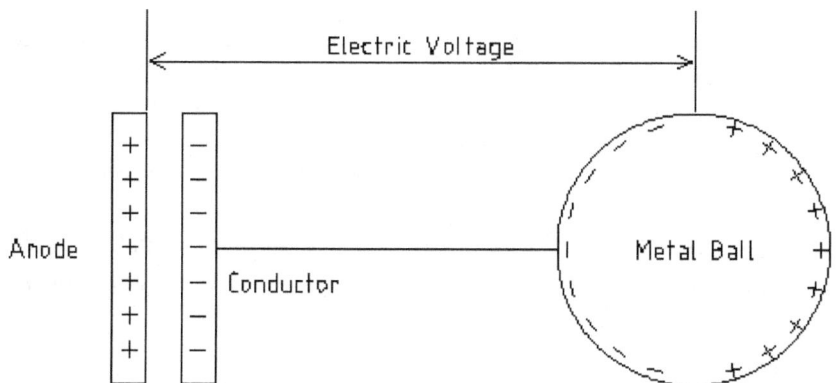

Illustration 8: Difference in Electric Potential between the ball and the Anode

The only thing missing from the entire process now, is the way we can use those extracted electrons. That can be done with the insertion of the conductive device as the coil wrapped around the conductor or the passive resistor directly connected to the conductor, so the electrons moving from the ball through the conductor then through the resistor can be harvested.

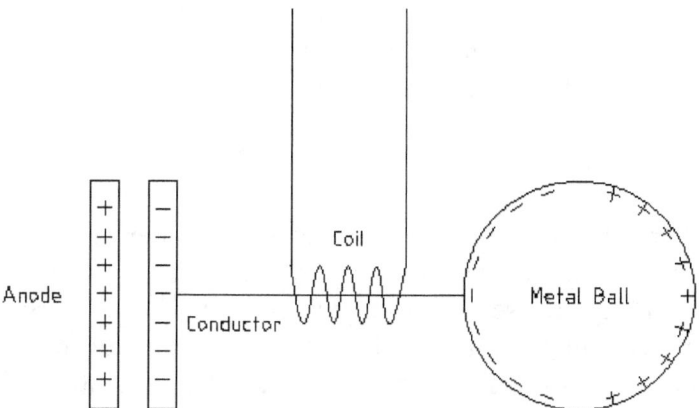

Illustration 9: Using the extracted free electrons to generate EMF in the coil

By wrapping the coil around the conductor, movement of electrons from the ball to the electrode is generating EMF in the coil therefore energy is generated in the coil. Electrons are then losing the speed and energy.

> **NOTE:** This is the version of electrons gathering, that I'm not sure it will work :)

This is supposed to work on the following principle:

"If the electricity is conducted through the middle of the enclosed circuit path, in that circuit path, emf (electromotive force) is generated."

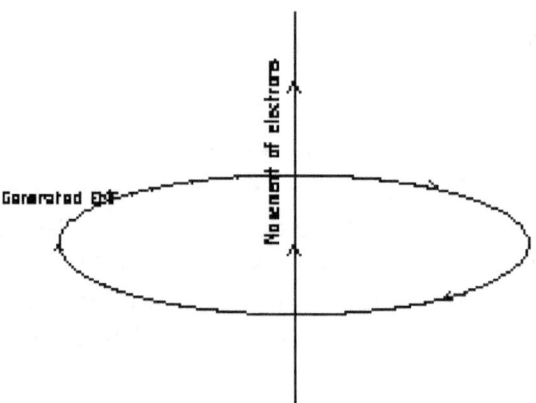

Illustration 10: Generation of Electromotive Force

The second way of free electrons gathering, is the way where an resistor is inserted on the conductor, so the electrons that are trying to escape the metal ball need to go through the resistor. By going through the resistor, electrons are transferring the energy that they get from the attraction power, to the heat in the resistor.

See the drawing on the next page.

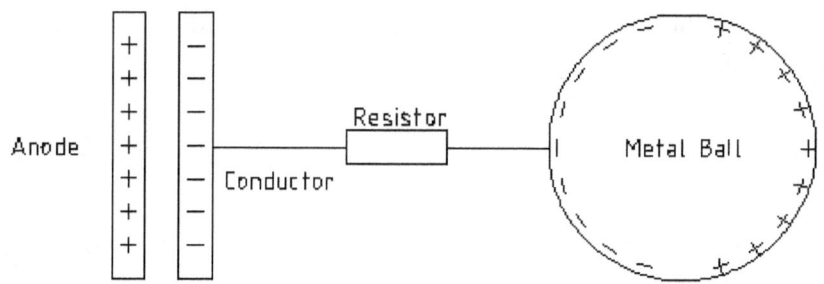

Illustration 11: Using the energy from the escaping free electrons to power the resistor and create heat.

This way, energy of the electrons can be fully used to generate heat in the resistor, the same way the heat generation works in the regular commercial electrical heating units.

However, the third way of gathering electrons is the most promising one. Rather than inserting passive electric device, we can instead place the charger for the batteries, therefore the electrons will not be passing through the conductor, but divert into the battery, where they will be fully extracted. This way we will completely remove the electrons from the circuit and will not risk the electron supersaturation of the electrode opposite to the anode.

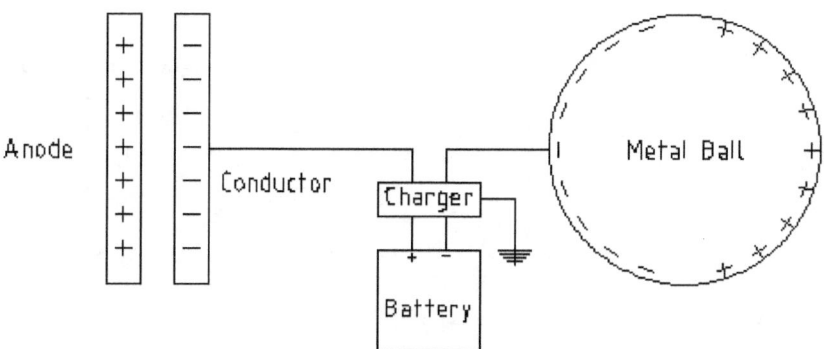

Illustration 12: Charging the battery with the Electron Extractor

When this extractor fully constructed, it would be consisted of following:

1. Metal object (Source of free electrons)
2. Anode (positive charged electrode used to attract free electrons)
3. Cathode (negative charged electrode used to repel free electrons)
4. Connector (connects an conductor with the metal object)
5. Conductor (connects metal object with the electrode opposite to the anode)
6. Electrode (stands opposite to the anode so it can help attract free electrons from distance)
7. Battery Charger (installed on the conductor so it can divert the flow of the extracted electrons)
8. Battery (stores extracted free electrons)
9. High Voltage DC Source (creates electric potential difference and charges anode and cathode)
10. Additional Conductors (for connecting the voltage source and the positive and negative charged electrodes)

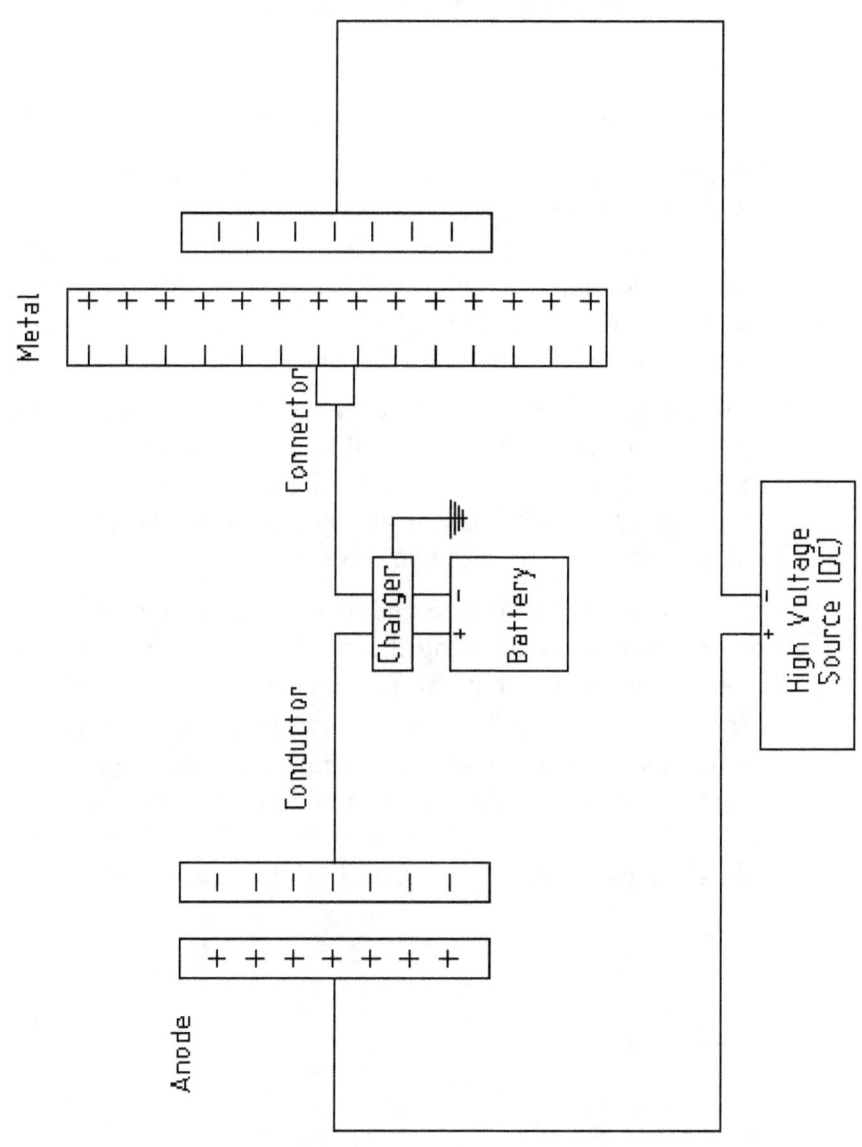

Chapter 3

Parts and 3D View

- Metal Object

    We already know that a cubic meter of aluminum contains $18.1*10^{28}$ of free electrons, or 8055388.1726 Ah of energy. So if you take any peace of aluminum, you can easily calculate its energy potential by dividing the volume of 1 m³ with the volume of the peace you have in mind, and you can calculate its energy accordingly.

    For example if you take a peace of aluminum, 0.1 m by 0.5 m by 0.5 m, its volume would be 0.025 m³, that is 1/40 of the 1m³ of aluminum. Now, if you divide 8055388.1726 with 40, you will get 201384.704315 Ah of energy, stored in this peace of aluminum.

    You can easily calculate the energy potential of every other material, by dividing the number of electrons in that material with the number of electrons in one coulomb ($1C= 6.241509*10^{18}$ e), you will then get the number of coulombs in that peace of metal. By multiplying the number of coulombs in that material with constant 0.00027777777778 (1C = 0.00027777777778 Ah), you will then get the energy potential of that peace of metal.

    Illustration 13: Peace of Aluminum 0.5*0.5*0.1

- Anode

    Anode is the positive charged electrode. Usually, there is current flow through the anode, however, in this case, we need to make this electrode charged with positive charge, but unmovable. To do this, we will connect the peace of metal with the positive DC voltage output, and that way we can create electron-attracting electrode.

    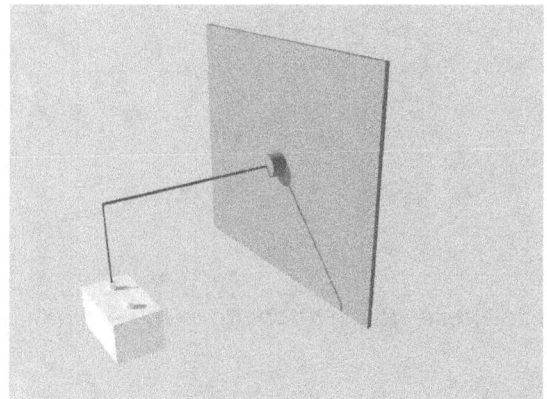
    Illustration 14: Copper plate connected to the positive DC Voltage Output

- Cathode is the basically same thing as the anode, except, this time, the copper plate is connected to the negative output of the DC Voltage Output.

- Connector is used to connect the conductor with the Free Electron Source. It eases the connection so when the metal peace is dried out, you can easily replace it.

- Conductor is used to connect the source of free electrons with the electrode opposite to the anode. On the conductor there can be installed extraction unit as battery charger.

- Electrode stands opposite to the anode, and it is basicaly metal plate where anode induces negative charge from the source of free electrons.

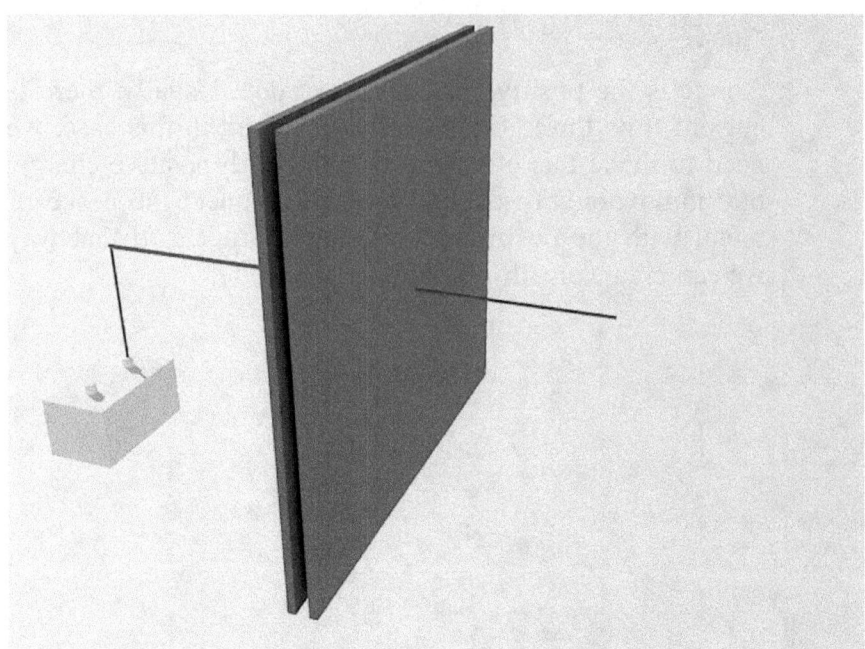
Illustration 15: An Electrode opposite to the Anode

- Battery Charger is the basic electronic device made to capture moving electrons (current) and to charge the battery with that current.
- Battery used in this circuit is the basic 45+ Ah battery, like one you can find in your car.
- High Voltage DC Source is used to charge the electrodes (anode and the cathode) so these electrodes can generate electrostatic field which attracts the electrons from the free electron source. This is not the power source, this is only used to charge the anode and the cathode. In that process, small amount of electricity is lost, but after electrodes are charged, there is no more power loss, there is only permanent electrostatic field in the electrodes.
- Various conductors are used to connect all of the circuit parts.

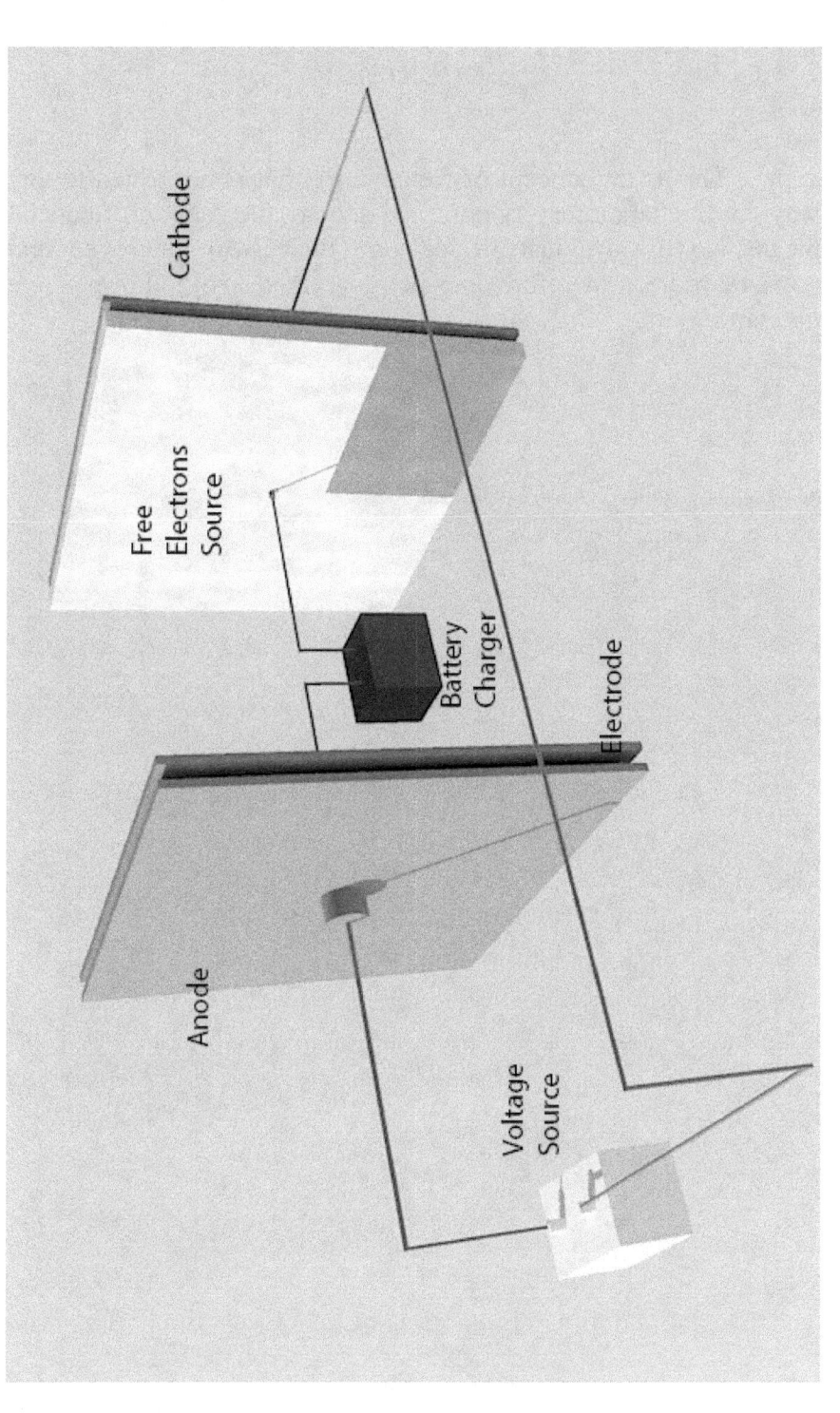

Conclusion

This is an concept of free energy that is not tested in any way in the laboratory conditions due to the lack of financial means, so it stays just an idea, an theoretical, unproved and untested project, therefore does not guarantee any kind of positive outcome.

www.ingramcontent.com/pod-product-compliance
Lightning Source LLC
Chambersburg PA
CBHW070735180526
45167CB00004B/1758